JOSEPH MIDTHUN   SAMUEL HITI

BUILDING BLOCKS OF SCIENCE

FORCE AND MOTION

WORLD BOOK

a Scott Fetzer company
Chicago

www.worldbookonline.com

World Book, Inc.
233 N. Michigan Avenue
Chicago, IL 60601
U.S.A.

For information about other World Book publications, visit our website at http://www.worldbookonline.com or call 1-800-WORLDBK (967-5325).

For information about sales to schools and libraries, call 1-800-975-3250 (United States); 1-800-837-5365 (Canada).

Library of Congress Cataloging-in-Publication Data

Force and motion.
     p. cm. -- (Building blocks of science)
   Includes index.
   Summary: "A graphic nonfiction volume that introduces the properties of force and motion. Features include several photographic pages, a glossary, additional resource list, and an index"-- Provided by publisher.
   ISBN 978-0-7166-1423-4
   1. Force and energy. 2. Motion. I. World Book, Inc.
   QC73.F67 2012
   531.6--dc23
                              2011025975

Building Blocks of Science
Set ISBN: 978-0-7166-1420-3

Printed in China by Leo Paper Products LTD., Heshan, Guangdong
1st printing December 2011

Acknowledgments:
Created by Samuel Hiti and Joseph Midthun.
Art by Samuel Hiti. Written by Joseph Midthun.

© David R. Frazier Photolibrary/Alamy Images 24;
© Leslie Garland Picture Library/Alamy Images 25;
NASA 9; © Shutterstock 8, 26, 27

ATTENTION, READER!

Some characters in this series throw large objects from tall buildings, play with fire, ride on bicycle handlebars, and perform other dangerous acts. However, they are CARTOON CHARACTERS. Please do not try any of these things at home because you could seriously harm yourself—or others around you!

STAFF

Executive Committee
President: Donald D. Keller
Vice President and Editor in Chief: Paul A. Kobasa
Vice President, Marketing/
   Digital Products: Sean Klunder
Vice President, International: Richard Flower
Director, Human Resources: Bev Ecker

Editorial
Associate Manager, Supplementary
   Publications: Cassie Mayer
Writer and Letterer: Joseph Midthun
Editors: Mike DuRoss and Brian Johnson
Researcher: Annie Brodsky
Manager, Contracts & Compliance
   (Rights & Permissions): Loranne K. Shields

Manufacturing/Pre-Press/Graphics and Design
Director: Carma Fazio
Manufacturing Manager: Steven Hueppchen
Production/Technology Manager:
   Anne Fritzinger
Proofreader: Emilie Schrage
Manager, Graphics and Design: Tom Evans
Coordinator, Design Development and
   Production: Brenda B. Tropinski
Book Design: Samuel Hiti
Photographs Editor: Kathy Creech

TABLE OF CONTENTS

There is a glossary on page 30. Terms defined in the glossary are in type **that looks like this** on their first appearance.

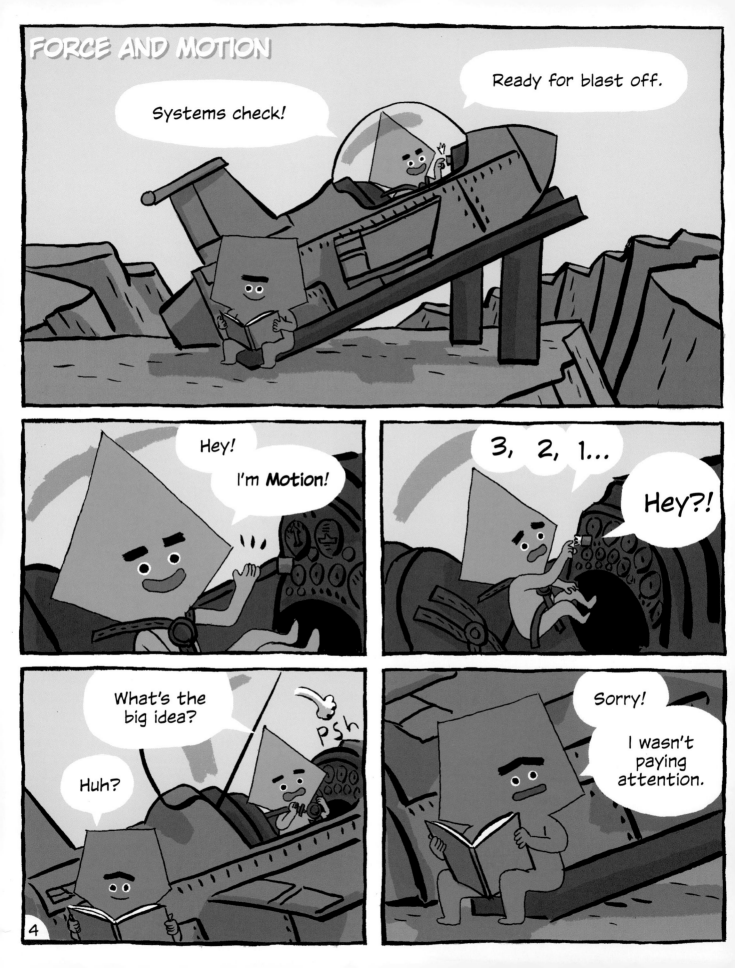

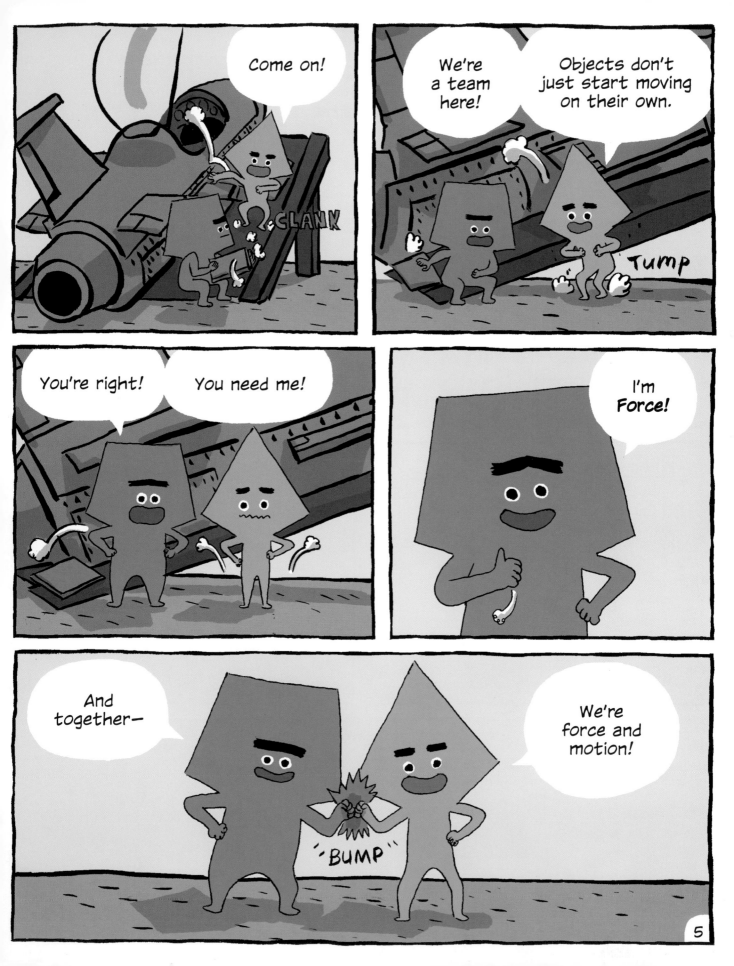

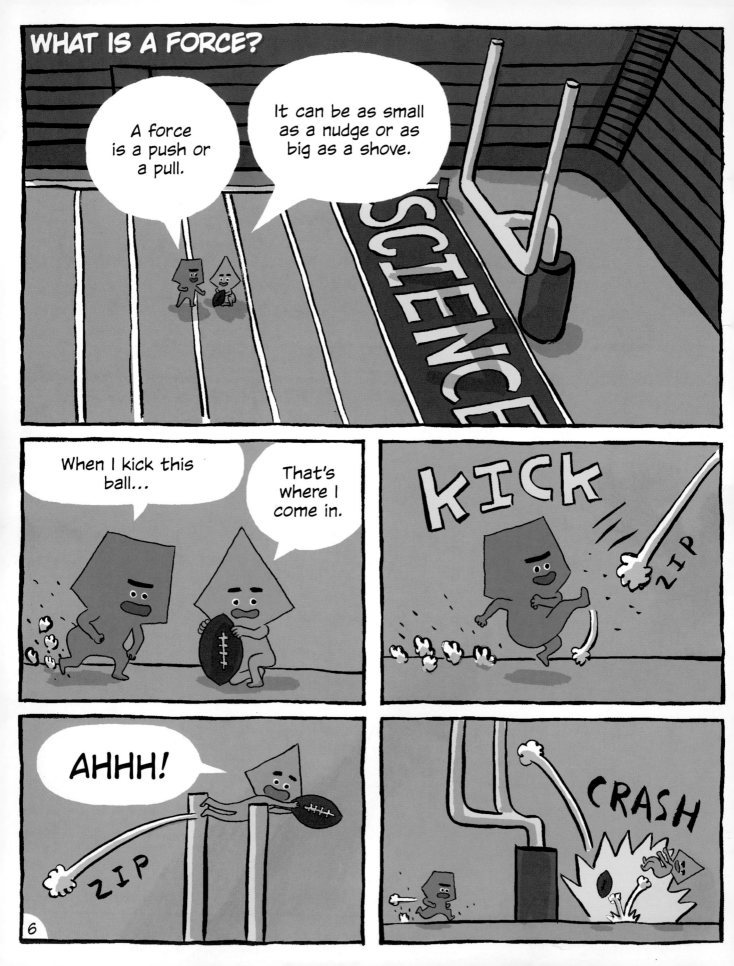

7

FORCES ALL AROUND US

There are many forces around you every day.

Look at this bulldozer!

That's a mechanical force at work.

That's a lot of force!

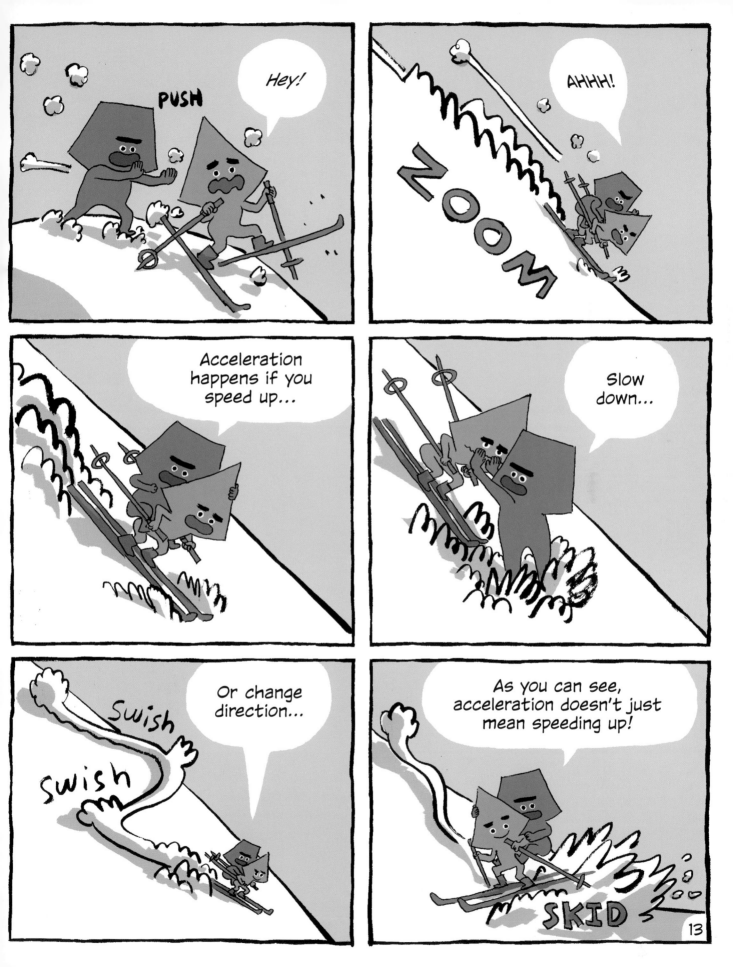

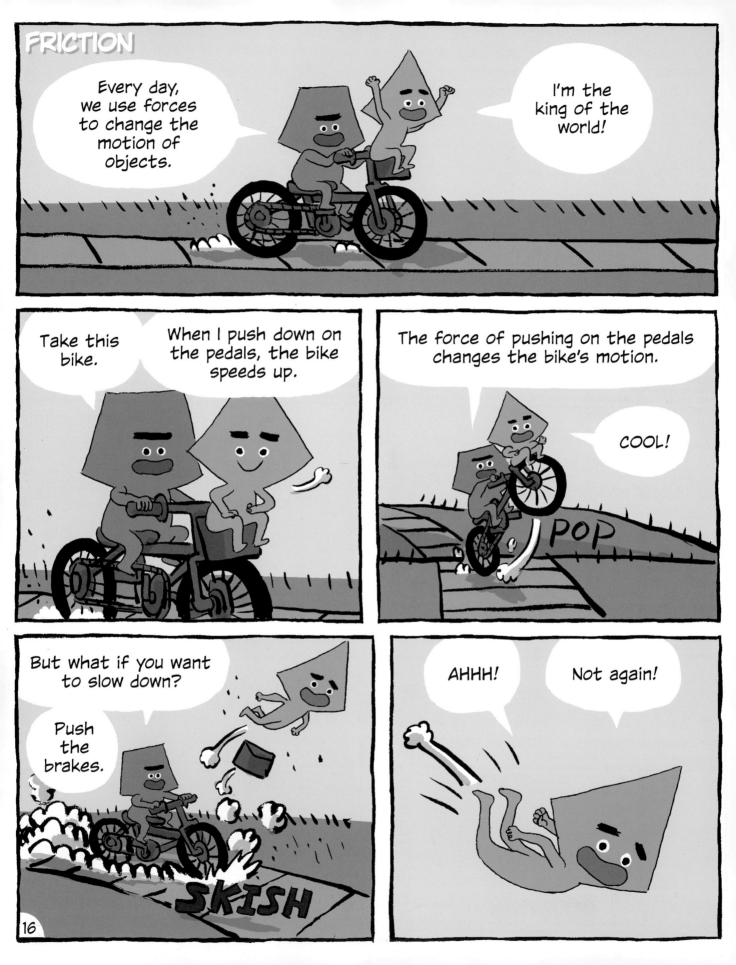

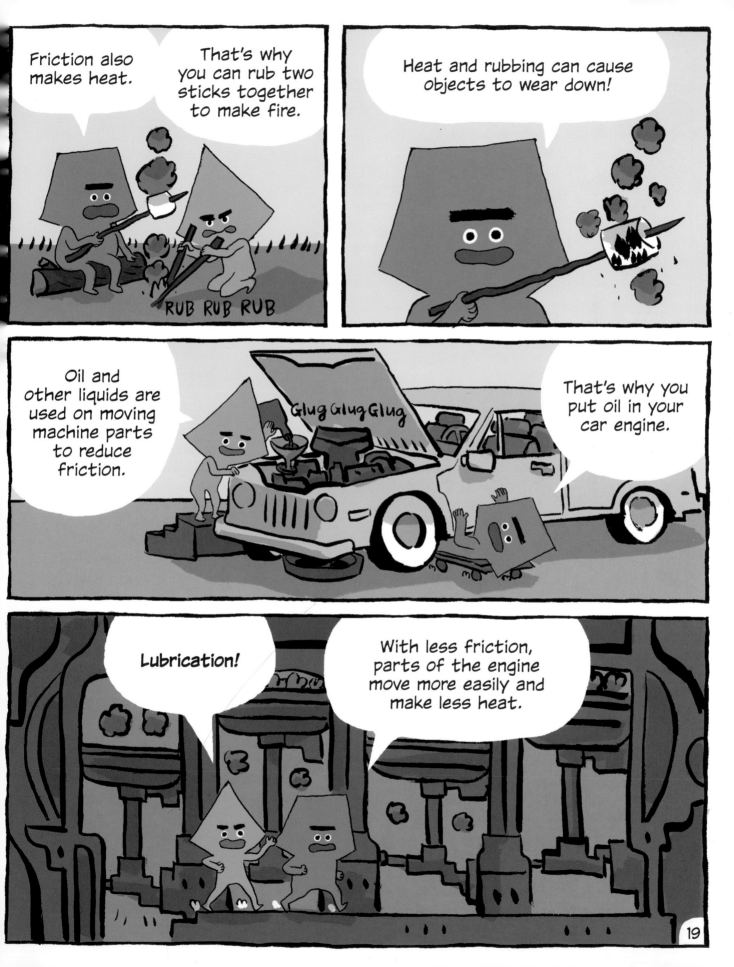

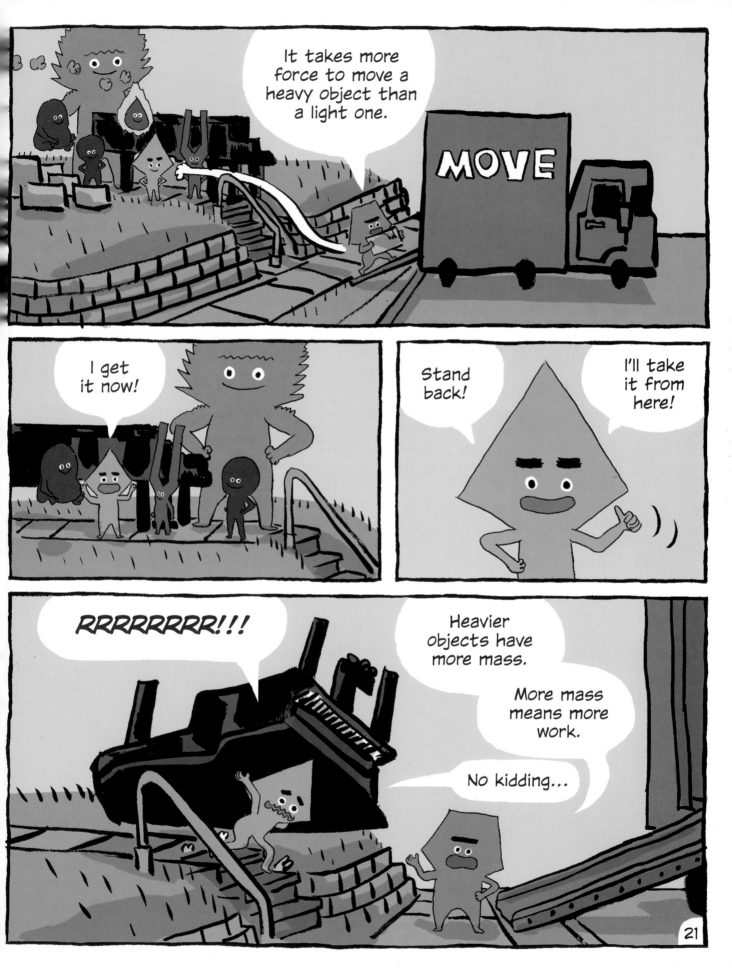

SIMPLE MACHINES

There are six kinds of simple machines.

An **inclined plane** is a flat, slanted surface. It can be used to raise heavy loads.

A **lever** is a bar that moves on a fixed point.

CLANK

scratch
scratch

What the...

A **screw** is an inclined plane wrapped into a spiral.

A **wedge** is two inclined planes back to back. A wedge is used to split materials.

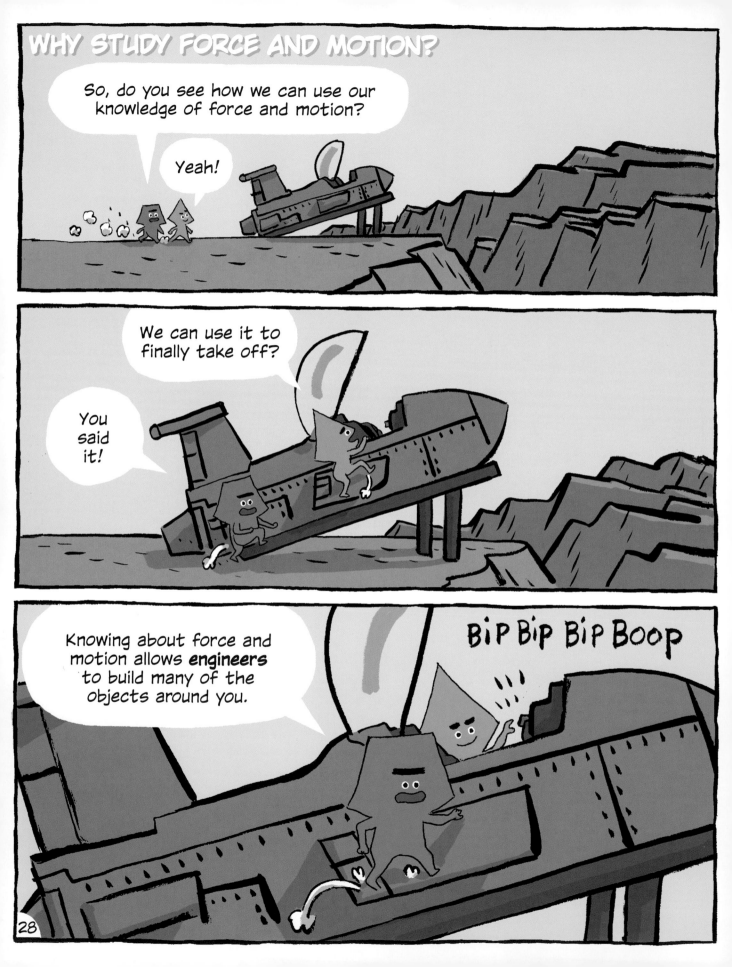

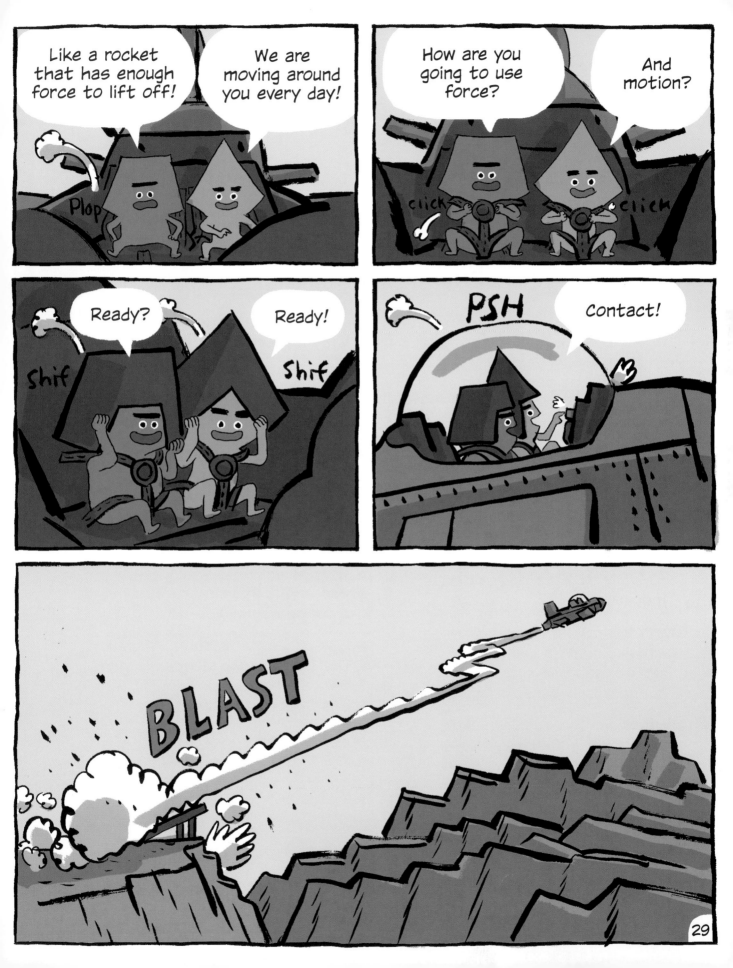

GLOSSARY

acceleration a change in the speed or direction of an object.

distance the amount of space between two points.

engineer a person who plans and builds engines, machines, roads, bridges, canals, forts, or the like.

force a push or a pull.

friction rubbing between objects that slows them down and produces heat.

gravity a force that attracts all objects toward one another.

inclined plane a simple machine shaped like a ramp.

inertia the tendency of objects to stay at rest or stay in motion.

lever a simple machine consisting of a rod or bar that rests and turns on a support called a fulcrum.

lubrication the act of making machinery smooth and easy to work by putting on oil or grease.

magnetism a force produced by the motion of electrons in a material.

mass the amount of matter in an object.

matter what all things are made of.

mechanical force the force applied when two objects touch each other.

motion a change in position.

physics the science that deals with matter and energy.

pulley a simple machine made of a rope or chain wrapped around a spinning wheel.

screw a simple machine shaped like a ramp wrapped around a central shaft.

simple machine any of six basic tools that change the way force is used to do work.

speed the distance traveled in a certain time.

wedge a simple machine shaped like two inclined planes placed back to back with an edge that cuts or slices.

wheel and axle a simple machine with a big wheel attached to a post.

FIND OUT MORE

Books

Can You Feel the Force? by Richard Hammond (DK Publishing, 2006)

A Crash Course in Forces and Motion with Max Axiom, Super Scientist by Emily Sohn (Capstone Press, 2007)

Experiments with Simple Machines by Salvatore Tocci (Children's Press, 2003)

How Fast Is It? A Zippy Book About Speed by Ben Hillman (Scholastic, 2008)

Inclined Planes and Wedges by Sally M. Walker and others (Lerner Publications, 2002)

Levers by Sally M. Walker and others (Lerner Publications, 2002)

Motion: Push and Pull, Fast and Slow by Darlene R. Stille and Sheree Boyd (Picture Window Books, 2004)

Science Experiments with Forces by Sally Nankivell-Aston and Dorothy Jackson (Franklin Watts, 2000)

Wheels and Axles by Sally M. Walker and others (Lerner Publications, 2002)

Websites

EdHeads: Simple Machines
http://www.edheads.org/activities/simple-machines/index.htm
The games on this interactive site explore the machines we use every day to do work.

Exploratorium: Science Snacks About Mechanics
http://www.exploratorium.edu/snacks/iconmechanics.html
Quick and easy experiments turn everyday objects into physics lessons at this website.

The Franklin Institute: Simple Machines
http://www.fi.edu/qa97/spotlight3/spotlight3.html
If you want more information about different machines and how they work, you can find it on this website from the Franklin Institute.

Inventor's Toolbox
http://www.mos.org/sln/Leonardo/InventorsToolbox.html
Explore simple machines—and the ways in which they combine to make more complex machines—at this educational site from the Museum of Science.

Physics4Kids: Motion
http://www.physics4kids.com/files/motion_intro.html
Want to learn more about the different kinds of forces at work in the world? Check out this physics site for kids.

Physics Games at PBS Kids
http://pbskids.org/games/physics.html
Familiar characters help you explore simple physics concepts at this fun and educational website from PBS.

Science Kids: Physics
http://www.sciencekids.co.nz/gamesactivities.html
Play around with friction and learn about gravity at this educational website.

INDEX